YOUR KNOWLEDGE HAS VALUE

- We will publish your bachelor's and master's thesis, essays and papers

- Your own eBook and book - sold worldwide in all relevant shops

- Earn money with each sale

Upload your text at www.GRIN.com and publish for free

Bibliographic information published by the German National Library:

The German National Library lists this publication in the National Bibliography; detailed bibliographic data are available on the Internet at http://dnb.dnb.de .

Imprint:

Copyright © 2017 GRIN Verlag, Open Publishing GmbH
Print and binding: Books on Demand GmbH, Norderstedt Germany
ISBN: 9783668540620

This book at GRIN:

http://www.grin.com/en/e-book/375896/anti-acne-activity-of-medicinal-and-herbal-sources

Prem Jose Vazhacharickal, John Joseph, Jiby John Mathew, Sajeshkumar N.K., E. S. Sumayya

Anti-acne activity of medicinal and herbal sources

GRIN Publishing

GRIN - Your knowledge has value

Since its foundation in 1998, GRIN has specialized in publishing academic texts by students, college teachers and other academics as e-book and printed book. The website www.grin.com is an ideal platform for presenting term papers, final papers, scientific essays, dissertations and specialist books.

Visit us on the internet:

http://www.grin.com/

http://www.facebook.com/grincom

http://www.twitter.com/grin_com

Anti-acne activity of medicinal and herbal sources: an overview

Prem Jose Vazhacharickal, John Joseph, Jiby John Mathew, Sajeshkumar N. K, and E. S Sumayya

ACKNOWLEDGEMENTS

Firstly, we thank God **Almighty whose** blessing were always with us and helped us to complete this project work successfully.

We wish to thank our beloved Manager **Rev. Fr. Dr. George Njarakunnel,** Respected Principal **Dr. Joseph V.J,** Vice Principal **Fr. Joseph Allencheril,** Bursar **Shaji Augustine** and the Management for providing all the necessary facilities in carrying out the study. We express our sincere thanks to **Mr. Binoy A Mulanthra** (lab in charge, Department of Biotechnology) for the support. This research work will not be possible with the co-operation of many farmers.

Lastly, we extend indebt thanks to patents, friends, and well-wishers for their love and support.

Prem Jose Vazhacharickal, John Joseph, Jiby John Mathew, Sajeshkumar N. K and E. S Sumayya

Table of contents

Table of figures

Table of tables

List of abbreviations

%	: Percentage
°C	: Degree celsius
µl	: Microlitre
ml	: Millilitre
mm	: Millimetre
sp	: Species
g	: Gram
NaCl	: Sodium chloride
MHA	: Mueller hinton agar
NA	: Nutrient agar
MA	: MacConkey agar
BA	: Blood agar
P. acne	: *Propionibacteium acne*
S epidermidis	: *Staphylococcus epidermidis*
S aureus	: *Staphylococcus aureus*
T. conoides	: *Turbinaria conoides*
DRESS	: Drug Reaction with Eosinophilia and Systemic Symptoms
CNS	: Central Nervous System
SIE	: *Selagineila involvens* extract

Anti-acne activity of medicinal and herbal sources: an overview

Prem Jose Vazhacharickal[1], John Joseph, Jiby John Mathew, Sajeshkumar N. K and E. S Sumayya

Abstract

Acne is the most common skin disorder with a very high prevalence. Acne may be treated with a combination of remedies including over – the counter skin care, and medications, and chemical or laser procedures. All these treatments are comparatively costlier. Hence in the current study, attempts are being made to search for new and cheaper remedy for acne. Here anti acne activity of various samples [Paste, cucumber, tomato, multani mitty (fullers earth) ,Guava leaf, turmeric and ointment(control)] was studied against *Staphylococcus sp.* The most effective samples used is guava leaf.

Keywords: Acne; Staphylococcus; Biochemical identification.

1. Introduction

Acne vulgaris is one of the commonest skin disorders, for which dermatologists are still struggling since years to treat successfully. It mainly affects adolescents, through it may present at any age. It is almost a universal disease occurring in all races and affecting 95% of 16-year-old boys and 83% of 16-year-old girls to some degree. The incidence of severity of acne, peak at 40% in 14-17-year-old girls and 35% in boys aged 16-19 year.

Acne by definition is a multi-factorial chronic inflammatory disease of pilosebaceous units. It affects the skin of the face, neck and upper trunk. These particular sebaceous follicles have capacious follicular channels and voluminous, multi-acninar sebaceous glands. Acne develops when these specialized follicles undergo pathogenic alteration that results in the formation of non-inflammatory lesions [comedons] and inflammatory lesions [papules, pustules and nodules].

Staphylococcus epidermidis and *Propionibacterium acne* are considered as the major skin bacteria that cause the formation of acne. *Propionibacterium acne*, a gram-positive an anaerobic pathogen, plays an imp role in the pathogenesis of acne.

It is implicated in the development of inflammatory acne by its capability to activate complements and by its ability to metabolize sebaceous triglycerides into fatty acids, which chemotactically attract neutrophils. *Staphylococcus epidermidis*, an aerobic organism, is usually involved in super facial infections within the sebaceous unit.

Modern acne therapy has been designed to interrupt the pathogenic pathway at one or more points. Tropical or Systemic therapy is available for the treatment of acne, which includes comedolytic agents and antibiotics and various anti-inflammatory drugs and Systemic therapy includes antibiotics, zinc and hormones. The excessive use of antibiotics for long period has led to increased resistance in acne causing bacteria; *Propionibacterium acne* and *Staphylococcus epidermidis* against a number of antibiotics used to treat acne.

The area's most vulnerable to acne are the largest oil secreting glands present in the face, back and trunk. It is a chronic inflammatory disease of multifactorial etiology affecting more than 85% of teenagers and frequently continues into adulthood.

The influencing factors of acne include excess sebum secretion, hyper keratinization of the hair follicle, oxidative stress and the release of inflammatory mediators. Acne is related to abnormalities in sebaceous gland function particularly in teenage. The hyper secretion of hormone androgen stimulates higher sebum secretion in sebaceous gland. The secreted sebum normally contains a mixture of lipids; squalene wax and cholesterol both in free and in ester forms and triglycerides that naturally provide a skin barrier function. However, the resulting abnormalities in sebaceous glands due to hormonal effects alter sebum composition and decrease linoleic acid content. Thus, the normal skin barrier function is impaired. In addition, the deficiency of linoleic acid in the follicle promotes the growth of normal flora like P. acnes. Proliferation of *Propionibacterium acne* leads to inflammatory lesions and severe acne.

In addition, *Staphylococcus epidermidis* and *Pitryosporum ovale* are present in acne lesions. Although the other microorganisms are present in acne lesions, the prevalent bacterium implicated in the clinical course of acne is *Propionibacterium acnes*. It is a gram positive, anaerobic bacterium that mostly resides in the pilosebaceous follicles of the skin.

Although P. acnes area member of the normal skin commensal, it plays a critical role in the development of acne [both inflammatory and non-inflammatory] when it becomes overgrown and colonizes the pilosebaceous unit.

Staphylococcus epidermidis, an aerobic organism, usually involves in superficial infections within the sebaceous unit. These factors provide a potential target for the treatment.

Propionibacterium acnes and *Staphylococcus epidermidis* are the target sites of antiacne drugs. Long term use of antibiotics against acne is outdated because of exacerbated antibiotic resistance. The development of antibiotic resistance is multifactorial, including the specific nature of the relationship of bacteria to antibiotics, how the antibacterial is used, host characteristics, and environmental factors. To overcome the problem of antibiotic resistance, medicinal plants have been extensively studied as alternative treatments for diseases.

1.1 Aim

The aim of this study is to find anti- acne activity of *Staphylococcus sp* against medicinal and herbal sources.

2. Review of literature

Propionibacterium acnes and *Staphylococcus epidermidis* have been recognized as pus-forming bacteria triggering an inflammation in acne (Chomnawang, 2005; Lertsatitthanakorn, 2008). *Propionibacterium acnes* is an anaerobic pathogen, which plays an important role in the pathogenesis of acne by inducing certain inflammatory mediators and comedogenesis (Jain and Sangal, 2009).

Propionibacterium acnes are found on the skin of neonates, but true colonization begins 1-3 years prior to sexual maturity. During this time, number of *Propionibacterium acnes* rise from fewer than $10/cm^2$ to about $106/cm^2$, chiefly on the face and upper thorax. In the lipid-rich microenvironment of the hair follicle, *Propionibacterium acnes* produce inflammatory mediators that result in papules, pustules and late rnodulocystic lesions that are typical of inflammatory acne *Propionibacterium acnes* increase sebum production and inflammation at the site of pilosebaceous follicles. It was reported that lipid mediators are able to interfere with sebocytes differentiation and sebogenesis of acne through the activation of pathways related to peroxisome proliferators-activated receptors (Halliwell and Gutteridge, 1998). *Propionibacterium acnes* is a relatively slow growing aero tolerant anaerobic gram-positive bacterium. This bacterium primarily lives on the fatty acid present in the sebaceous glands or sebum secreted by follicles. It has the ability to produce propionic acid and catalase in presence of indole, nitrate or both in dole and nitrate. When a pore is blocked by the secretion, the anaerobic bacterium starts over growing and secretes chemical which break down the wall of the pore and leads to acne lesion. Propionibacterium resembles Corynebacterium in morphology and arrangement.

Since *Propionibacterium acnes* is a common resident of the pilosebaceous glands of the human skin and acne is caused in part from an infection, it was suppressed with topical and oral antibiotics. Antibiotics were primarily used in treatment against acne for more than 40 years (Tzellos, 2011).

Normally used topical and oral antibiotics are Clindamycin, Erythromycin, Triclosan, Tetracycline, Minocycline and Metronidazole. The most frequently used topical antibiotic for acne was Clindamycin which is available as a solution, lotion or gel at 1% strength. Salicylic acid is also used in treatment of acne as it was found to be one of the good cleansers that has both anti ¬inflammatory and mild comedolytic effects. It is used in treatment of mild acne or as an adjunctive agent (David, 2003). Physical treatment in the form of lesion removal, photo-therapy is also helpful in few of them (Leyden, 1997). Now a day's new therapeutic modalities and various permutation and combinations have been designed in topical agents; that include benzoyl peroxide, antibiotics, Retinoids etc. These combinations when tried on patients increase the frequency and severity of dryness, scaling, erythema, burning, stinging and itching (Draelos, 2010). Minocycline has an increased risk of severe adverse effects compared to other tetracyclines. It may induce hypersensitivity reactions affecting the liver, lungs, kidneys or multiple organs (Drug Reaction with Eosinophilia and Systemic Symptoms [DRESS] syndrome) in the first weeks of treatment and with long-term treatment, may cause autoimmune reactions (Systemic Lupus Erythematosus, autoimmune Hepatitis). In addition, Central Nervous System (CNS) symptoms such as dizziness are more frequent compared to other tetracyclines. Long-term treatment may also induce hyperpigmentation of the skin.

Therapy with these medications generally lasts from several weeks to months, but may sometimes even continue for years. In particular, this long-term exposure to oral antibiotics creates tremendous selective pressures for the emergence of resistant strains of *Propionibacterium acnes* (Ross, 2001).

In fact, resistance to *Propionibacterium acnes* develops in 50% of individuals following treatment with both topical and oral antibiotics (Esperson, 1998). The rate of *Propionibacterium acnes* resistance corresponds to the length of treatment (Tan, 2005). In a prospective study of 151 patients, the rate of resistance in patients who had never been on antibiotics was 0, compared to 6.25% in patients on short-term (6-18 weeks) antibiotics and 21.6% in patients on longer periods of antibiotics.

Since the use of antibiotics has resulted in resistant strains of *Propionibacterium acnes* and also the many side effects associated with its prolonged use resulted in people opting for holistic treatment. Medicinal plants are known to have enormous therapeutic capabilities that modern medicine is searching for. Thus, it became an alternative therapy for consumers, which is cost effective when compared to modern treatment involving antibiotics.

Herbs are safe, efficacious and multifunctional. The ingredients in topical acne treatments, particularly herbs and naturally derived compounds, have received considerable interest as they have fewer adverse effects than synthetic agents (Keren, 2003)

The anti-bacterial and anti-fungal activity of four seaweeds namely *Sargassum binderi*, Amphiroasp., *Turbinaria conoides* and *Halimeda macroloba* Decaisne from the east coast of Gulf of Thailand was studied. The test organisms used in the study were *Staphylococcus aureus*, S. epidermidis, *Propionibacterium acnes*, *Proteus mirabilis* and *Candida albicans.* it was found that aqueous and ethanolic extracts of all the four species were found to have antimicrobial activity. However aqueous extract of *T. Conoides* demonstrated the maximum activity. Therefore, aqueous extract of *T. Conoides* was further evaluated for its anti-inflammatory effect using EPP-induced ear edema and carrageenin-induced hind paw edema tests. Results revealed anti-inflammatory activity of aqueous extract of *T. Conoides* comparable to that of phenylebutazol and acetylsalicylic acid (Walailuck, 2011). Studies on anti-oxidant and anti-biotic activity of *Selagineila involvens* extract (SIE) revealed to have an antioxidant effect in a dose-dependent manner in the hydroxyl radical-mediated oxidation test. Thus, SIE could be used as a safe non-antibiotic that can be used to treat acne. It was also known to reduce the non-specific initiation and augmentation phase of the inflammatory response of *Propionibacterium acnes* (Joo, 2008).

Herbal plants and its parts are effective alternatives used to treat acne. Herbal medicines are recommended since these are powerful cleansers that could clear your body of unwanted harmful toxins and enrich the body with useful nutrient minerals.

Ayurveda, which is one of the oldest system of traditional medicine of India is now regaining its pace, the herbs in Ayurveda used to treat acne are Neem, Tea tree oil, Sandal wood, Burdock root, Red clover, *Aloe vera*, Calendula and Lavender. All these plants are well known for their anti-microbial activity. *Calendula officinalis*,

Cassia tora and *Momordica charantia* are well known herbs used in ayurvedic traditional medicine for their effectiveness against wide range of diseases including skin infections as they have diverse secondary metabolites that is responsible for their antibacterial activity (Roopashree and Raman, 2008).

In a study, the antimicrobial activity of Indian medicinal plants against *Propionibacterium acnes* that caused acne vulgaris was evaluated. Extracts of *Rauwolfia serpentine* (roots), *Piper nigrum* (seeds), *Azadirachta indica* (leaves), *Ficus religiosa* (leaves), *Euphorbia hirta* (roots), *Ocimum sanctum* (leaves), *Phyllanthus niruri* (whole plants), *Cardiospermum halicacabum* (leaves), *Mordica charantia* (fruits), *Casaurina equisetifolia* (fruits), *Cynodon dactylon* (leaves) and *Jasmimum sambac* (flowers) were tested for antimicrobial activity by agar well diffusion and broth dilution method. The results from the disc diffusion method showed that 07 medicinal plants could inhibit the growth of *Propionibacterium acnes.* Among these *Azadirachta indica, Momordica charantia, Casuarina equisetifolia, Rauwolfia serpentina, Cardiospermum halicacabum, Phyllanthus niruri* and *Piper nigrum* had strong inhibitory effects. Based on broth dilution method, *Rauwolfia serpentina* extract had the greatest antimicrobial effect. *Piper nigrum* and *Momordica charantia* showed outstanding antimicrobial properties against *Propionibacterium acnes* based on the agar well diffusion assay. In bioautography assay, the *Rauwolfia serpentine* extract produced strong inhibition zones against *Propionibacterium acnes.* Phytochemical screening of *Rauwolfia serpentine* revealed the presence of alkaloids (berberine and harmane) which could possibly be responsible (Harisaranraj, 2010) for inhibition.

Medicinal plants belonging to families Liliaceae, Rutaceae, Zingiberaceae, Myrtaceae, Lamiaceae etc. contains alkaloids, tannins, flavonoids, terpenoids, volatile oil and essential oil which are reported to have significant effect against acne causing bacterial (Deepak, 2011).

A stable anti-acne preparation was developed in a study by determining the main chemical constituents of volatile oil from leaves of *Eucalyptus globules* and *Psidium guajava.* Volatile oils from these plants leaves were extracted and analysed for their main chemical components. Antimicrobial activity of the oils was determined by agar diffusion and micro dilution method. These volatile oils showed good anti-microbial activity against *Propionibacterium acnes*. The main constituents present in the volatile oils of Eucalyptus and Guava leaves were gamma-terpentine and alpha-pinene. These results suggested that the preparations incorporating the volatile oils of Eucalyptus and Guava could be used in anti-acne formulation (Sirivan, 2008).

Bioactivity studies on Java citronella oil (*Cymbopogon winterianus*) revealed good anti-acne activity with potential for use in the development of topical anti-acne preparations. *Jasmimum auriculatuma* small herb of the family Oleaceae found growing wild in South India and the Western peninsula was reported to have antioxidant and antibacterial activities. The alcohol free defatted extract of *Jasmimum auriculatum* leaves has been reported to contain the essential oils lupeol and Jasminol that has a potent anti-acne activity.

Lemon grass oil contained citral as the major constituent and it showed a good anti-acne activity. Thus, suggesting that lemon grass can be used to treat acne vulgaris (Faiyazuddin, 2010). The dried herbs assessed were *Chrysanthemum morifolium* Juhua), *Lonicera japonica* (honeysuckle), *Jasmimum sambac* (jasmine), *Lavandulaan gustifolia* (lavender), *Rosa damascene* (rose), *Osmanthus fragrans* (osmanthus), *Eucommia ulmoides* (duzhong), *Gynostemma pentaphyllum* (jiaogulan), *Cymbopogon citrates* (lemon grass) and *Ilex paraguariensis* (yerba mate). The herbs were ground and extracted with methanol in a series of steps. They were tested against *Propionibacterium acnes* by determining Minimum inhibitory concentration (MIC).

Many herbal plants were screened for acne treatment; the plants included Ocimum, *Tabernaemontana dinaricata*, *Melaleuca alternifolia* (Tea tree) and Aloe vera. These were being made in the form of topical formulations. In-vitro antibacterial studies were performed against *Propionibacterium acnes*. This was carried out by well diffusion method and Tetracycline was used as standard. Ocimum was developed in hydroalcoholic extract, *Tabernaemontana dinaricata*, Tea tree and *Aloe vera* was developed in ethanolic extract. *Ocimum sanctum* and Tea tree showed the greater zone of inhibition (Sawarkar, 2010).

The antibacterial activity of oriental herbal extracts of *Angelica dahurica* and *Glycyrrhiza glabra* was studied against *Propionibacterium acnes*. *Glycyrrhiza glabra* gave more promising results and therefore could be helpful in prevention and treatment of acne lesions. Some plant extracts exhibited both antimicrobial and anti-inflammatory effects against *Propionibacterium acnes* (Nam, (2003).

Three herbal extracts namely *Rosa damascene*, *Eucommia ulmoides* and *Ilex paraguariensis* were used in a study with 3 concentrations (0.5, 1 and 2 mg/mL). Results of the studies showed that both *Eucommia ulmoides* and *Ilex paraguariensis* possessed antimicrobial and anti-inflammatory effects against *Propionibacterium acnes*.

Pitikamardhini an herbal formulation, which consists of plants such as *Rubia cordifolia*, *Symplocos racemosa*, *Acorus calamus*, *Coriandrum sativum* and *Citrus limonis* was effective in curing Acne vulgaris. A clinical trial was made which gave good results and also improved facial complexion. Present studies revealed that Pitikamardhini could be used in clinical treatment of acne (Burade, 2009).

The antibacterial activity of plant extract against *Propionibacterium acnes* and S. Epidermidis was studied by cup plate method. In this method, the macerated fruit and fruit decoction of *Terminalia chebula* and *Terminalia bellarica* (belonging to Combretaceae family) was used individually and in combination. The combination of 20% gave the best zone of inhibition (around 26 mm) (Sawarkar, 2011). *Terminalia bellarica* was studied for its anti-microbial activity against *Staphylococcus aureus*, *Streptococcus pneumoniae*, *Salmonella typhi* and many others by crude and methanolic extracts of the seed by disc diffusion method. The zone of inhibition was around 15.5 to 28.0 mm for crude extract and 14.0 to 30.0 mm for methanolic extract. The study suggested that methanolic extract was better than the crude one. The results also indicate that *T. Bellarica* dry fruit possesses potential broad spectrum antimicrobial activity (Elizabeth, 2005).

3. Hypothesis

The current research work is based on the following hypothesis

1) Herbal extracts show a promising effect on acne treatment compared to chemical ointments.

2) These herbal extracts differ in their anti-acne activities.

4. Materials and Methods

4.1 Study area

Kerala state covers an area of 38,863 km^2 with a population density of 859 per km^2 and spread across 14 districts. The climate is characterized by tropical wet and dry with average annual rainfall amounts to 2,817 ± 406 mm and mean annual temperature is 26.8°C (averages from 1871-2005; Krishnakumar et al., 2009). Maximum rainfall occurs from June to September mainly due to South West Monsoon and temperatures are highest in May and November.

4.2 Collection of samples

The sample containing organism is collected from the pimples, of persons affected by Swabbing method and transferred to peptone water.

4.3 Pure culture preparation

Bacterial pure culture was prepared by streak plate method. One loop full of enrichment culture from the flasks was streaked on nutrient agar plates. The growth of the bacterial colonies was measured after 24-48 hr incubation at 28°C. Morphologically dissimilar colonies were randomly selected and sub cultured onto nutrient medium and maintained at 4°C for bacterial characterization.

4.4 Morphological and biochemical tests

Identification of the isolates were performed according to their morphological, cultural and biochemical characteristics by following Bergey's Manual of Systematic Bacteriology. All the isolates were subjected to Gram staining and specific biochemical tests.

4.4.1 Colony morphology

This was done to determine the morphology of selected strains on the basis of shape, size and colour.

4.4.2 Gram staining

A clean grease free slide was taken and a smear of the bacterial culture was made on it with a sterile loop. The smear was air-dried and then heat fixed. Then it was subjected to the following staining reagents:

(i) Flooded with Crystal violet for 1 min. followed by washing with running distilled water.
(ii) Again, flooded with Gram's Iodine for 1 min. followed by washing with running distilled water.
(iii) Then the slide was flooded with Gram's Decolourizer for 30 sec.
(iv) After that the slide was counter stained with Safranin for 30 sec, followed by washing with running distilled water.
(v) The slide was air dried and cell morphology was checked under microscope.

4.4.3 Motility test

This test is done to check the motility of the microorganism. The isolates were taken in a cavity slide from a 24 hr inoculated broth and observe under the microscope.

4.4.4 Sugar fermentation test

Various carbohydrates were used and added to peptone water to test the fermentation reaction. Gas production was detected by placing an inverted Durham's tube in the medium. The carbohydrates used include sucrose, lactose, and glucose. The result was read by the change in the colour from blue to yellow.

4.4.5 Oxidative fermentation test

The test is used to detect the oxidative fermentation characteristics of the test organisms. Certain bacteria use carbohydrates oxidatively and produce acids only in open tubes and contain other produced acids both in open and closed tubes fermentative. Positive organisms showed colour change from blue to yellow.

4.4.6 Nitrate reduction test

It is used to study the ability of bacteria to reduce nitrate to nitrite. The organism that possess the enzyme nitrate reductase reduce nitrate to nitrite. Positive organism showed a pink colour when alpha naphthalamine and sulphanilic acid was added to the medium.

4.4.7 Coagulase test

Coagulase test determine *Staphylococcus aureus* from other members in the genus. *S. aureus* produce an enzyme coagulase which activates prothrombin to clot rabbit plasma. The clot formation in the tube helps to identify the organism as *Staphylococcus aureus*.

4.4.8 Oxidase test

Oxidase test is used to detect the oxidase producing organisms. The cytochrome oxidase enzyme is able to oxidize the substrate tetra methyl paraphenylenediamenedihydrochloride forming a coloured product. Dark purple end product will be visible when a small amount of culture rubbed on the substrate impregnated on the paper.

4.4.9 Catalase test

It helps to demonstrate whether the bacteria are capable of producing catalase enzyme. Effervescence formation indicates positive result.

4.4.10 Mannitol motility test

The test is used to detect the ability of the test organism to ferment mannnitol and to check the motility of the organism. Positive organisms showed a change in the colour of the medium from red to yellow after incubation. Motile organisms showed diffused growth in the medium.

4.4.11 Indole production test

This test is conducted to distinguish the bacteria based on the ability to produce indole from tryptophan. Indole broth contains tryptophan rich with peptone and sodium chloride (NaCl). Formation of red ring within seconds on addition of Kovac's reagent shows a positive result.

4.4.12 Methyl red test

The test is used to study the ability of bacteria to produce sufficient acid during fermentation of glucose. Positive result shows a red colour on addition of methyl red indicator.

4.4.13 Voges-Proskauer test

VP test is conducted by to study the production of acetone during fermentative degradation of glucose. Positive organisms showed formation of pink colour on addition of VP agent.

4.4.14 Citrate utilization test

This test is used to study the ability of bacteria to utilize citrate as the sole source of carbon. Development of blue colour after 24 – 48 hours in the tube inoculated with positive organisms shows positive result.

4.4.15 Urease test

This test was used to detect the ability of bacteria to produce enzyme urease which hydrolyses urea and releases ammonia and carbon dioxide. Ammonia reacts in solution to form ammonium carbonate, which is alkaline and lead to the increase in pH. Phenol red changed its colour from yellow to red in alkaline pH thus indicating the presence of urease activity.

4.5 Extract preparation

The extracts used are Turmeric and guava leaf, Tomato, Cucumber, Paste, Multani mitty and Ointment [control] in 1:1 ratio. The extracts were prepared aseptically. The samples were cut into pieces with a sterile scalpel and using mortar and pestle it was crushed and the extract was taken with muslin cloth and separated. It was then mixed with sterile distilled water in 1:1 ratio. The extracts of multani mjtty, ointment ad paste was taken approximately about 1g and mixed with sterile distilled water.

4.6 Antimicrobial activity

Mueller-Hinton Agar was prepared and sterilized. After sterilization, the media was allowed to hand bearable temperature and poured into the Petri plates and kept for solidification. Wells are prepared in each of the plates using well puncture. The bacterial cultures were seeded over the media. After swabbing, the extracts of Turmeric and Guava leaf, Ointment [control], Paste, Cucumber, Tomato and Multani mitty (fullers earth) were loaded into the wells and kept for incubation at 37°C for 24 hours. After incubation, the zone of inhibition was measured using a zone measurement scale.

4.7 Treatment

Check the presence of zone of inhibition, around the well. Then to the sterilized Mueller-Hinton Agar, well is prepared by using Well Puncture. After seeding bacteria, extracts that show zone of inhibition [turmeric and guava leaf], is loaded to the wells with varying concentrations [50 µl,100 µl ,150 µl and 200 µl]. Incubate at 37°C for 24hours. After incubation; zone of inhibition was measured using a zone measurement scale.

4.8 Statistical analysis

The survey results were analysed and descriptive statistics were done using SPSS 12.0 (SPSS Inc., an IBM Company, Chicago, USA) and graphs were generated using Sigma Plot 7 (Systat Software Inc., Chicago, USA).

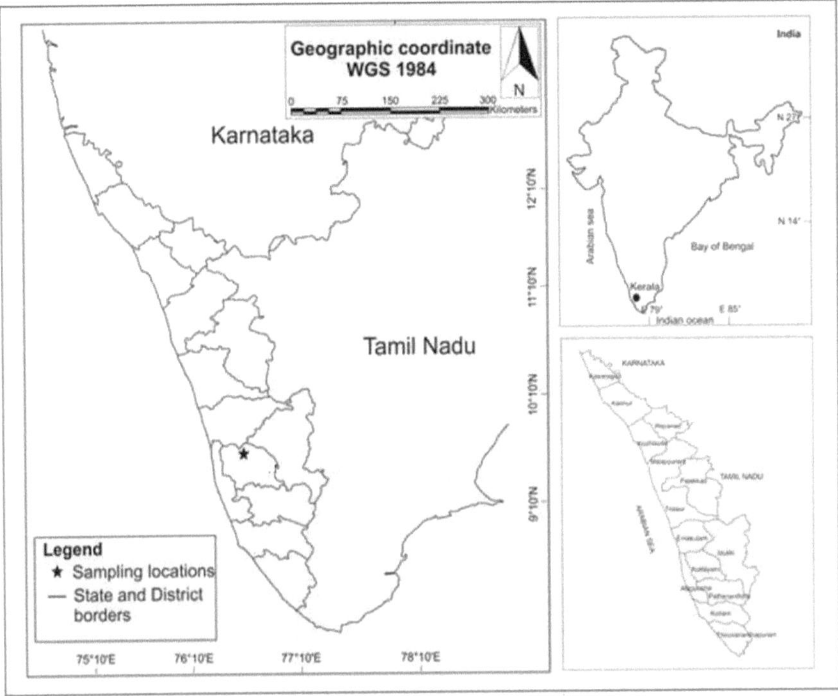

Figure 1. Map of Kerala showing the sample collection point. Authors own work.

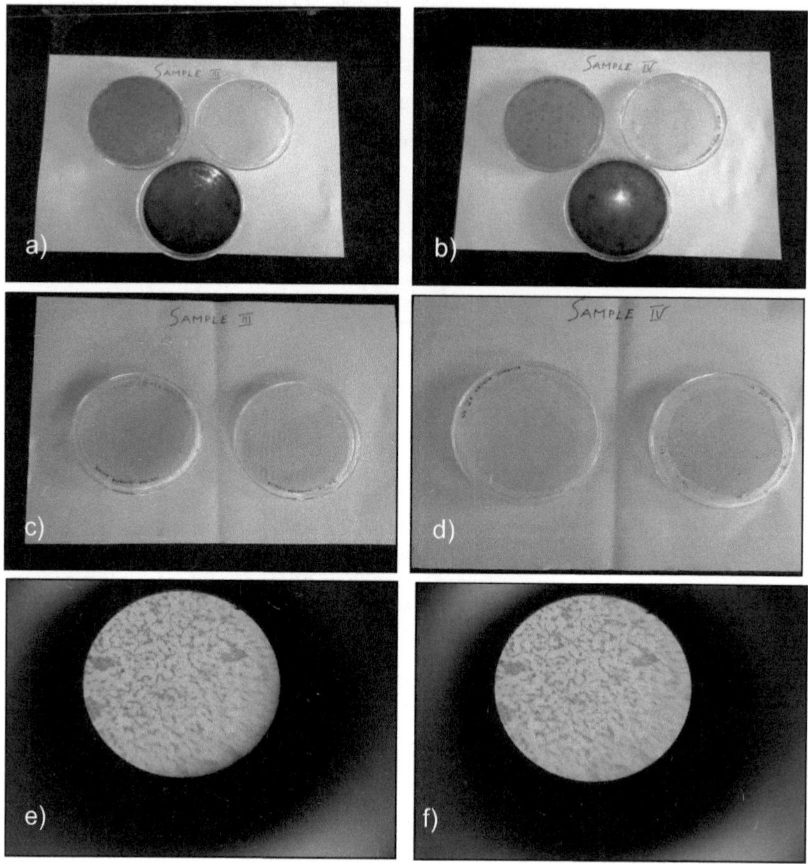

Figure 2. Details of the isolated *Staphylococcus sp.* a), b), c) and d) growth of the organism on various culture media, e) and f) microscopic examinations. Authors own images.

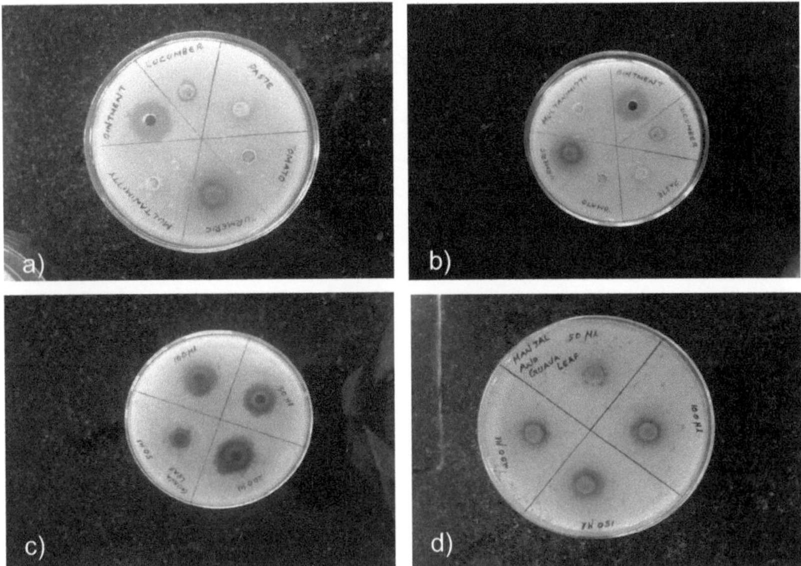

Figure 3. Details of antimicrobial activity of various extracts against *Staphylococcus sp.* and zone of inhibition a) concentration 50 µl, b) concentration 100 µl, c) concentration 150 µl, d) concentration 200 µl. Authors own images.

Table 1. Various biochemical tests of the isolated bacterial strains (isolate 1; *Staphylococcus sp.* and 2; *Pseudomonas sp.*).

Tests	Bacterial strains	
	Isolate 1	*Isolate 2*
Carbohydrate test		
• Glucose	Positive	Negative
• Lactose	Positive	Negative
• Sucrose	Positive	Negative
• Maltose	Positive	Negative
Nitrate test	Positive	Negative
Coagulase test	Positive	Negative
Oxidase test	Negative	Negative
Catalase test	Positive	Negative
Mannitol Motility	Fermentative, non motile	Non fermentative, motile
Indole test	Negative	Negative
Methyl red test	Positive	Negative
Voges-Proskauer	Positive	Negative
Citrate	Positive	Positive
Urease	Positive	Negative
Gram staining	Positive	
Motility	Non motile	Motile

5. Results and discussion

There are many skin care products that can be used to treat pimples, but there are also many home remedies that can use to get rid of the pimples. Acne vulgaris (Simply acne) is a common human skin disease, characterized by areas of seborrhoea (Scaly red skin), comedones (blackheads and white heads), papules (pinheads), nodules (large papules), pimples and possibly scarring. Aside from scarring its main effects are physiological. Many different treatments exist for acne including benzoyl peroxide, antibiotics, retinoid, antiseborrheic medications, anti-androgen medications, hormonal treatments, salicylic acid, alpha hydroxyl acid, azelaic acid, nicotinamide and keratolytic soaps. They are believed to work in at least four different ways, including the following: normalizing shedding and sebum production into the pore to prevent blockage, killing *Staphylococcus sp.* and *Propionibacterium acne.* Anti-inflammatory effects and hormonal manipulations. Numerous natural products have been investigated for treating people with acne. In the study Paste, cucumber, tomato, multani mitty, Guava leaf, turmeric and ointment (control) where taken and tested for screening antimicrobial activities against *Staphylococcus sp* and *Propionibacterium acne.* It has been found that majority of the isolates were subjected to antimicrobial action.

Table 2. Antimicrobial activity of various extracts against *Staphylococcus sp.* and zone of inhibition (mm).

Trial No.	Loading volume (μl)	Zone of inhibition (mm)					
		Guava leaf	Turmeric	Guava leaf + Turmeric	Cucumber	Ointment	Paste
1	50	10.2 ± 0.26	4.3 ± 0.27	15.7 ± 0.11	-	20.7 ± 0.11	15.1 ± 0.11
2	100	20.4 ± 0.16	8.1 ± 0.33	20.3 ± 0.22	-	20.8 ± 0.11	15.3 ± 0.11
3	150	18.6 ± 0.22	12.4 ± 0.38	25.2 ± 0.41	10.3 ± 0.11	20.5 ± 0.11	15.3 ± 0.11
4	200	20.1 ± 0.27	10.6 ± 0.20	25.6 ± 0.20	10.2 ± 0.10	20.5 ± 0.11	15.2 ± 0.11

Numbers represent means ± one standard deviation (SD) of the mean

6. Conclusions

Acne vulgaris is a chronic inflammatory disorder of pilosebaceous units, its prevalence is highest in adolescence. *Propionibacterium acnes* and *Staphylococcus epidermidis* have been recognized as pus forming bacteria triggering inflammation in acne. Medicinal and herbal treatments are most commonly used for anti-acne treatment. The organism present on the acne have been isolated and then the isolate has been cultured. For this study, several extracts [Paste, cucumber, Tomato, multani mitty, Guava leaf, turmeric and ointment (control)] was prepared. Then their anti- microbial activity was done. The antimicrobial activity of *Staphylococcus sp.* was performed at varying concentrations. The result revealed that the guava leaf is most resistant. This study will aim the clinician to prescribe best medicinal and herbal anti-acne treatment.

References

Abdoul-Latif, F., Edou, P., Mohamed, N., Ali, A., Djama, S., Obame, L. C., & Dicko, M. (2010). Antimicrobial and antioxidant activities of essential oil and methanol extract of *Jasminum sambac* from Djibouti. *African Journal of Plant Science*, 4(3), 38-43.

Bhaskar, G., Arshia, S., & Priyadarshini, S. R. B. (2009). Formulation and evaluation of topical polyherbal antiacne gels containing *Garcinia mangostana* and *Aloe vera*. *Pharmacognosy Magazine*, 5(19), 93.

Boonchum, W., Peerapornpisal, Y., Kanjanapothi, D., Pekkoh, J., Amornlerdpison, D., Pumas, C., & Vacharapiyasophon, P. (2011). Antimicrobial and anti-inflammatory properties of various seaweeds from the Gulf of Thailand. *International Journal of Agriculture and Biology*, 13(1), 100-104.

Chaudhary, S. (2010). Antiacne activity of some Indian herbal drugs. *International Journal of Pharma Professional's Research*, 1(1), 78-80.

Chomnawang, M. T., Surassmo, S., Nukoolkarn, V. S., & Gritsanapan, W. (2005). Antimicrobial effects of Thai medicinal plants against acne-inducing bacteria. *Journal of Ethnopharmacology*, 101(1), 330-333.

David, C., & Liao, M. D. (2003). Management of acne. *The Journal of Family Practice*, 52, 44-51.

Deshpande, S. M., & Upadhyay, R. R. (1967). Chemical studies of Jasminum auriculatum (VAHL) leaves.

Draelos, Z. D., Potts, A., & Alió, S. A. (2010). Randomized tolerability analysis of clindamycin phosphate 1.2%-tretinoin 0.025% gel used with benzoyl peroxide wash 4% for acne vulgaris. *Cutis*, 86(6), 310-318.

Espersen, F. (1998). Resistance to antibiotics used in dermatological practice. *British Journal of Dermatology-Supplement*, 139, 4-8.

Faiyazuddin, M., Ali, J., Ahmad, S., Ahmad, N., Akhtar, J., & Baboota, S. (2010). Chromatographic analysis of trans and cis-citral in lemongrass oil and in a topical phytonanocosmeceutical formulation, and validation of the method. *JPC-Journal of Planar Chromatography-Modern TLC*, 23(3), 233-236.

Finney-Brown, T. (2010). Antimicrobial and anti-inflammatory activity against *P. acnes*. *Australian Journal of Medical Herbalism*, 22(2), 62-64.

Fu, Y., Chen, L., Zu, Y., Liu, Z., Liu, X., Liu, Y., & Efferth, T. (2009). The antibacterial activity of clove essential oil against Propionibacterium acnes and its mechanism of action. *Archives of Dermatology, 145*(1), 86-88.

Gutteridge, J., & Halliwell, B. (2000). Free radicals and antioxidants in the year 2000: a historical look to the future. *Annals of the New York Academy of Sciences, 899*(1), 136-147.

Harisaranraj, R. S., Babu, S., & Suresh, K. (2010). Antimicrobial properties of selected Indian medicinal plants against acne-inducing bacteria. *Ethnobotanical Leaflets, 14*(1), 84-94.

Jain, A., Sangal, L., Basal, E., Kaushal, G., & Agarwal, S. K. (2002). Anti-inflammatory effects of Erythromycin and Tetracycline on Propionibacterium. acnes induced production of chemotactic factors and reactive oxygen species by human neutrophils. *Dermatology Online Journal, 8*(2), 2-8.

Joo, S. S., Jang, S. K., Kim, S. G., Choi, J. S., Hwang, K. W., & Lee, D. I. (2008). Anti-acne activity of *Selaginella involvens* extract and its non-antibiotic antimicrobial potential on Propionibacterium acnes. *Phytotherapy Research, 22*(3), 335-339.

Kanlayavattanakul, M., & Lourith, N. (2011). Therapeutic agents and herbs in topical application for acne treatment. *International Journal of Cosmetic Science, 33*(4), 289-297.

Kim, S. S., Baik, J. S., Oh, T. H., Yoon, W. J., Lee, N. H., & Hyun, C. G. (2008). Biological activities of Korean *Citrus obovoides* and *Citrus natsudaidai* essential oils against acne-inducing bacteria. *Bioscience, Biotechnology, and Biochemistry, 72*(10), 2507-2513.

Krishnakumar, K. N., Rao, G. P., & Gopakumar, C. S. (2009). Rainfall trends in twentieth century over Kerala, India. Atmospheric environment, 43(11), 1940-1944.

Lee, T. W., Kim, J. C., & Hwang, S. J. (2003). Hydrogel patches containing triclosan for acne treatment. *European Journal of Pharmaceutics and Biopharmaceutics, 56*(3), 407-412.

Leyden, J. J. (1997). Therapy for acne vulgaris. *New England Journal of Medicine, 336*(16), 1156-1162.

Martin, K. W., & Ernst, E. (2003). Herbal medicines for treatment of bacterial infections: a review of controlled clinical trials. *Journal of Antimicrobial Chemotherapy*, *51*(2), 241-246.

Ochsendorf, F. (2010). Minocycline in acne vulgaris. *American Journal of Clinical Dermatology*, *11*(5), 327-341.

Pothitirat, W., Chomnawang, M. T., Supabphol, R., & Gritsanapan, W. (2009). Comparison of bioactive compounds content, free radical scavenging and anti-acne inducing bacteria activities of extracts from the mangosteen fruit rind at two stages of maturity. *Fitoterapia*, *80*(7), 442-447.

Ren, K. (2001). *U.S. Patent No. 6,183,747.* Washington, DC: U.S. Patent and Trademark Office.

Roopashree, T. S., Dang, R., Rani, R. S., & Narendra, C. (2008). Antibacterial activity of antipsoriatic herbs: *Cassia tora*, *Momordica charantia* and *Calendula officinalis*. *International Journal of Applied Research in Natural Products*, *1*(3), 20-28.

Ross, J. I., Snelling, A. M., Eady, E. A., Cove, J. H., Cunliffe, W. J., Leyden, J. J., & Oshima, S. (2001). Phenotypic and genotypic characterization of antibiotic-resistant *Propionibacterium acnes* isolated from acne patients attending dermatology clinics in Europe, the USA, Japan and Australia. *British Journal of Dermatology*, *144*(2), 339-346.

Sawarkar, H. A., Khadabadi, S. S., Mankar, D. M., Farooqui, I. A., & Jagtap, N. S. (2010). Development and biological evaluation of herbal anti-acne gel. *International Journal of PharmTech Research*, *2*(3), 2028-2031.

Shivanand, P., Nilam, M., & Viral, D. (2010). Herbs play an important role in the field of cosmetics. *International Journal of PharmTech Research*, *2*(1), 632-639.

Singh, D., Hatwar, B., & Nayak, S. (2011). Herbal plants and Propionibacterium acnes: An overview. Int. J. Biomed. Res, 2, 486-498.

Spencer, E. H., Ferdowsian, H. R., & Barnard, N. D. (2009). Diet and acne: a review of the evidence. *International Journal of Dermatology*, *48*(4), 339-347.

Tan, A. W., & Tan, H. H. (2005). Acne vulgaris: a review of antibiotic therapy. *Expert Opinion on Pharmacotherapy*, *6*(3), 409-418.

Tzellos, T., Zampeli, V., Makrantonaki, E., & Zouboulis, C. C. (2011). Treating acne with antibiotic-resistant bacterial colonization. *Expert Opinion on Pharmacotherapy, 12*(8), 1233-1247.

Vijayalakshmi, A., Tripura, A., & Ravichandiran, V. (2011). Development and evaluation of anti-acne products from *Terminalia arjuna* bark. *International Journal of ChemTech Research, 3*(1), 320-327.